A Bibliography for Flora Zambesiaca

A guide to general floristic works which should be
consulted by contributors to Flora Zambesiaca

G.V. Pope & R.K. Brummitt

Royal Botanic Gardens, Kew

Copyright 1991 © Royal Botanic Gardens, Kew

ISBN 0 947643 32 X

Typeset by Electronic Images, Cambridge

Printed by Target Litho, Cottenham, Cambridge

Contents

Introduction

General floristic works relevant to Flora Zambesiaca are listed below as a guide to contributors to the Flora. Ecological and phytogeographical works have been omitted unless they include significant species lists.

Treatments of single families or smaller taxonomic groups are not listed here but should as a matter of course be consulted and cited for each taxon dealt with in a Flora account.

A list of abbreviations for journals and other general works used in Flora Zambesiaca is given in a separate document.

Part 1   lists works dealing wholly or partly with one or more of the countries; Caprivi Strip, Botswana, Zambia, Zimbabwe, Malawi and Mozambique.

References are grouped by country and there alphabetically arranged by author. The Flora Zambesiaca form of citation and the Kew Library catalogue number are indicated (F = floras, T = Travels Room, q = quarto, p = pamphlets, other refs. = journals).

All plant names appearing in these works should wherever possible be accounted for in the Flora as either accepted names or as synonyms, but it may not be obligatory to cite every reference – see note on asterisked works below.

Part 2   lists works dealing with countries adjacent to the Flora Zambesiaca area, or with the African flora as a whole, which must be cited in the Flora wherever relevant.

Part 3   lists abbreviated references in chronological order.   The country concerned in each work is indicated. 'EXT'   refers to extra-Flora Zambesiaca literature.

* – Works which should be cited

References marked '*' in which a relevant name appears must be cited in a Flora Zambesiaca account.

Works not marked '*' may be cited or not according to the author's wishes.

There are, of course, many other works relating to areas outside the Flora Zambesiaca area but not listed here, which may be additionally cited if an author wishes. The present list is only of works which *must* be cited if relevant.

Acknowlegements

We are indebted to Drs. Jorge Paiva, James Seyani and Charlie Riches for supplying references, for Mozambique, Malawi and Botswana respectively, additional to those in our original list.

PART 1
Floristic works of the Flora Zambesiaca area

BOTSWANA

Allen A. A preliminary reconnaissance of the vegetation of Orapa and environs.
Botswana Notes and Records 10: 169–185 (1978).

   cite as:- **Allen in Botswana Notes and Records 10:   (1978).**

* **Barnes J.E. & Turton L.M.** A list of the Flowering Plants of Botswana in the Herbaria at the National Museum, Sebele and University of Botswana.
Botswana Soc. Gaborone. 1986.   **[F10.75]**

   cite as:- **Barnes & Turton, List Fl. Pl. Botswana at Nat. Mus., Sebele & Univ. Botswana: (1986).**

Blomberg-Ermatinger V. & Turton L. Some Flowering Plants of South Eastern Botswana.
Botswana Soc. Gaborone. 1988.

   cite as:- **Blomb.-Ermat. & Turton, Some Fl. Pl. SE. Botswana :   (1988).**

Bremekamp C.E.B. & Obermeyer A.A. Scientific results of the Vernay-Lang Kalahari Expedition, 1930: Sertum Kalahariense, a list of the plants collected.
Ann. Transv. Mus. 16: 399–455 (1935).   **[12.8]**

   cite as:- **Bremek. & Oberm. in Ann. Transv. Mus. 16:   (1935).**

Brown N.E. List of the plants collected in Ngamiland and the northern part of the Kalahari Desert, chiefly in the neighbourhood of Kwebe and along the Botletle and Lake rivers.
Bull. Misc. Inf., Kew 1909: 89–146 (1909).   **[1.2/95]**

   cite as:- **N.E. Br. in Bull. Misc. Inf., Kew 1909:   (1909).**

Cole M.M. & Brown R.C. The vegetation of the Ghanzi area of Western Botswana.
Journal of Biogeography 3: 169–196 (1976).

   cite as:- **Cole & Brown in Journ. Biogeogr. 3:   (1976).**

Grignon I. & Johnsen P. Towards a Check List of the Vascular Plants of Botswana.
Aarhus Univ. Denmark. 1986.   **[F10.75]**

   cite as:- **Grignon & Johnsen, Towards Check List Vasc. Pl. Botswana: (1986).**

Leistner O.A. Preliminary list of plants found in the Kalahari Gemsbok National Park.
Koedoe 2: 152–172 (1959).   **[F10.75]**

   cite as:- **Leistner in Koedoe 2:   (1959).**

Leistner O.A.   The plant ecology of the Southern Khalahari.
Bot. Surv. S. Afr. Mem. 38, (1967). map.   [12/20]

   cite as:- **Leistner in Bot. Surv. S. Afr. Mem. 38:   1967).**

Miller O.B.   Check-list of the Forest Trees and Shrubs of the British Empire,
No. 6, Bechuanaland Protectorate (1948).
The Scrivener Press. Oxford. 1948.   [F10.75]

   cite as:- **O.B. Mill., Check-list For. Trees Shrubs Bech. Prot.:   (1948).**

*   **Miller O.B.**   The woody plants of the Bechuanaland Protectorate.
Journ. S. Afr. Bot. 18: 1–100 (1952); 19: 177–182 (1953).   [12/18]

   cite as:- **O.B. Mill. in Journ. S. Afr. Bot. [vol.]:   [year].**

Mott P.J.   Key to the common flowering plants in Botswana.
Univ. of Botswana Biology Dept., Government Printer, Gaborone (1976).

   cite as:- **Mott, Key Common Fl. Pl. Bot.:   (1976).**

Passarge S.   Beobachtete und Gesammelte Pflanzen der Kalahariregion.
Die Kalahari [Anhang 9]: 785–795 (1904).   [qT 10.75]

   cite as:- **Passarge, Die Kalahari: (1904).**

Pole Evans I.B.   A reconnaissance trip through the eastern part of the
Bechuanaland Protectorate, April 1931, and an expedition to Ngamiland, 1937.
Bot. Surv. S. Afr. Mem. 21, (1948). map.   [12/20]

   cite as:- **Pole Evans in Bot. Surv. S. Afr. Mem. 21:   (1948).**

Simpson C.D.   A detailed vegetation study on the Chobe River in North East
Botswana.
Kirkia 10: 185–227 (1975).   [10.74/1]

   cite as:- **Simpson in Kirkia 10:   (1975).**

Woollard J.   A Vegetative Key to the Plants of South Eastern Botswana.
Univ. College of Botswana. Gaborone. 1981.

   cite as:- **Woollard, Vegetative Key Woody Pl. SE. Botswana: (1981).**

MALAWI

Banda E.A.K. & Seyani J.H. Vegetation Survey of Sanjika Hill.
University of Malawi, Herbarium. 1, (1974); 2, (1975); 3, (1975). **[pq F10.72]**

cite as:- **Banda & Seyani, Veg. Surv. Sanjika Hill [vol.]:** **[year].**

Banda E.A.K. & Patel I.H. Vegetation Survey of Chilunga Estate,
Chancellor College, Zomba.
[University of Malawi, Herbarium]. 1975. **[pq F10.72]**

cite as:-**Banda & Patel, Veg. Surv. Chilunga Estate, Chancellor College,
Zomba:** **(1975).**

Banda E.A.K. & Salubeni A.J. The Vegetation Survey of Kamuzu Academy
and Mtunthama, Kasungu.
[University of Malawi, Herbarium]. 1982. **[pq F10.72]**

cite as:- **Banda & Salubeni, Veg. Surv. Kamuzu Academy Mtunthama,
Kasungu:** **(1982).**

Banda E.A.K. & Morris B. Common Weeds of Malawi.
University of Malawi, Zomba. 1986. **[F10.72]**

cite as:- **Banda & Morris, Common Weeds Malawi:** **(1986).**

Berrie A. Pteridophyta collected in Malawi, with a preliminary checklist of
the Orders Psilotales and Lycopodiales.
Luso 2,1: 33–48 (1981). **[p 230 (10.72)]**

cite as:- **Berrie in Luso 2,1:** **(1981).**

Berrie A. A checklist of the Pteridophytes of Zomba Mountain, Malawi
Luso 5,2: 67–86 (1984). **[p 230 (10.72)]**

cite as:- **Berrie in Luso 5,2:** **(1984).**

* **Binns B.** A First Check List of the Herbaceous Flora of Malawi.
Govt. Printer, Zomba. 1968. **[F10.72]**

cite as:- **Binns, First Check List Herb. Fl. Malawi:** **(1968).**

Binns B. Dictionary of Plant Names in Malawi.
Govt. Printer, Zomba. 1972. **[F10.72]**

cite as:- **Binns, Dict. Pl. Names Malawi:** **(1972).**

Blackmore S., Dudley C.O. & Osborne P.L. An annotated check list of the
aquatic macrophytes of the Shire River, Malawi, with reference to potential
aquatic weeds.
Kirkia 13: 125–142 (1989). **[10.74/1]**

cite as:- **Blackmore et al. in Kirkia 13:** **(1989).**

Brass L.J. Vegetation of Nyasaland. Report on the Vernay Nyasaland
Expedition of 1946.
Mem. N.Y. Bot. Gard. 8,3: 161–190 (1953). **[F10.72]**

cite as:- **Brass in Mem. N.Y. Bot. Gard. 8,3:** **(1953).**

* **Brenan J.P.M. & collaborators.** Plants Collected by the Vernay Nyasaland Expedition of 1946.
Mem. N.Y. Bot. Gard. 8,3: 191–256 (1953); 8,5: 409–510 (1954); 9,1: 1–132 (1954).　　　　[F10.72]

　cite as:- [Author] in Mem. N.Y. Bot. Gard. [vol.], [part]:　　　[year].

Britten J., Baker E.G., Rendle A.B., Gepp A. & others. The plants of Milanje, Nyasaland, collected by Mr. Alexander Whyte FLS.
Trans. Linn. Soc. II, Bot. 4: 1–67 (1894).　　　　[1.2/99]

　cite as:- [Author] in Trans. Linn. Soc., ser. 2, Bot. 4:　　　(1894).

Brummitt R.K. Systematic list of Nyika botanical collections.
Wye College 1972 Malawi Project Final Report: 47–77 (1973).　　　[F10.72]

　cite as:- Brummitt in Wye Coll. Malawi Proj. Rep.:　　　(1973).

Burkill I.H. List of the known plants occurring in British Central Africa, and the British territory north of the Zambezi.
in H.H. Johnston, British Central Africa: 233–284 (1897).
[The Serpa Pinto specimens listed were all from Angola].　　　[F10.72]

　cite as:- Burkill in Johnston, Brit. Cent. Afr.:　　　(1897).

Burtt Davy J. & Hoyle A.C. Check-lists of the Forest Trees and Shrubs of the British Empire No. 2, Nyasaland Protectorate.
Imperial Forestry Institute, Oxford. 1936.　　　　[F10.72]

　cite as:- Burtt Davy & Hoyle, Check-lists For. Trees & Shrubs Brit. Emp. 2, Nyasaland:　　　(1936).

Chapman J.D. The Indigenous Conifers of Nyasaland.
The Hetherwick Press, Blantyre Mission, Malawi. 1957.

　cite as:- Chapman, Indig. Conifers Nyasal.:　　　(1957).

Chapman J.D. The Vegetation of the Mlanje Mountains, Nyasaland.
Govt. Printer, Zomba. 1962.　　　　[F10.72]

　cite as:- Chapman, Veg. Mlanje Mt.:　　　(1962).

Chapman J.D. Mpita Nkhalango – a lowland forest relic unique in Malawi.
Nyala 12: 3–26 (1988).　　　　[p F10.7]

　cite as:- Chapman in Nyala 12:　　　(1988).

* **Chapman J.D. & White F.** The Evergreen Forests of Malawi.
Commonwealth Forestry Institute, Oxford. 1970.　　　　[F10.72]

　cite as:- Chapman & White, Evergreen For. Malawi:　　　(1970).

Dowsett-Lemaire F. The forest vegetation of Mt. Mulanje (Malawi): a floristic and chorological study along an altidudinal gradient (650–1950 m.).
Bull. Jard. Bot. Nat. Belg. 58: 77–107 (1988).　　　　[1.5/11]

　cite as:- Dowsett-Lemaire in Bull. Jard. Bot. Nat. Belg. 58:　　　(1988).

Hall-Martin A.J. & Drummond R. B. Annotated list of plants collected in Lengwe National Park, Malawi.
Kirkia 12: 151–181 (1980).　　　　[10.74/1]

　cite as:- Hall-Martin & Drumm. in Kirkia 12:　　　(1980).

Howard-Williams C. A check list of the vascular plants of Lake Chilwa Malawi, with special reference to the influence of enviornmental factors on the distribution of taxa.
Kirkia 10: 563–579 (1977).   **[10.74/1]**

    cite as:- **Howard-Williams in Kirkia 10:**   **(1977).**

Moriarty A. Wild Flowers of Malawi.
Purnell, Cape Town & London. 1975.   **[F10.72]**

    cite as:- **Moriarty, Wild Fl. Malawi:**   **(1975).**

Pullinger J.S. & Kitchin A.M. Trees of Malawi.
Blantyre Print & Publishing, Blantyre. 1982.   **[F10.72]**

    cite as:- **Pullinger & Kitchin, Trees Malawi:**   **(1982).**

Seyani J.H., Salubeni A.J., Banda E.A.K. & Patel I.H. The Plant Species of the Lilongwe Nature Sanctuary.
National Herbarium of Malawi, Zomba. 1987.

    cite as:- **Seyani et al., Plant Sp. Lilongwe Nat. Sanctuary:**   **(1987).**

Sherry B.Y. & Ridgeway A.J. A Field Guide to Lengwe National Park.
Montfort Press, Limbe, Malawi. 1984.   **[F10.72]**

    cite as:- **Sherry & Ridgeway, Field Guide Lengwe Nat. Park:**   **(1984).**

Shorter C. Introduction to the Common Trees of Malawi.
Wildlife Society of Malawi. 1989.

    cite as:- **Shorter, Introd. Common Trees Malawi:**   **(1989).**

*   **Topham P.** Check List of the Forest Trees and Shrubs of the Nyasaland Protectorate. [second ed. of Burtt Davy & Hoyle, 1936].
Govt. Printer, Zomba. 1958.   **[F10.72]**

    cite as:- **Topham, Check List For. Trees Shrubs Nyasaland Prot.:**   **(1958).**

Wiehe P.O. Catalogue of Flowering Plants in the Herbarium of the Division of Plant Pathology, Zomba.
Dept. Agric., Zomba.   **[F10.72]**

    cite as:- **Wiehe, Cat. Fl. Pl. Herb. Pl. Path. Zomba:**   **(1952).**

Willan R.G.M. & McQueen D.R. Exotic Trees of the British Commonwealth: Nyasaland.

    cite as:- **Willan & McQueen, Exotic Trees Nyasaland:**   **(1957).**

Williamson J. Useful Plants of Malawi.
Univ. of Malawi. 1975.
[revised and extended edition of Useful Plants of Nyasaland 1956, Govt. Printer, Zomba].   **[630.9 (10.72)]**

    cite as:- **Williamson, Useful Pl. Malawi:**   **(1975).**

MOZAMBIQUE

Amico A. & Bavazzano R. Contributo alla conoscenza della flora della Zambesia inferiore (Mozambico).
Webbia 23: 247–303 (1968).   **[1.54/37]**

cite as:- **Amico & Bavazzano in Webbia 23:**   **(1968).**

Balsinhas A.A. The weeds of abandoned cotton fields in Mozambique.
Bothalia 14: 971–975 (1983).   **[12/9]**

cite as:- **Balsinhas in Bothalia 14:**   **(1983).**

Bertoloni G. Illustrazione di Piante Mozambicesi. Dissertazioni 1–4.
[Memoria Accad. Sci. Ist. Bologna, vols. 2–5] (1850–1855).   **[q F10.71]**

cite as:- **Bertoloni, Ill. Piante Mozamb. [Diss.]:**   **[year].**

Gomes e Sousa A.F. Subsídios para o Estudo da Flora do Niassa Português.
[Reprinted with new pagination from Bol. Soc. Estudos Colon. Moçamb.
No. 26] (1935).   **[F10.71]**

cite as:- **Gomes e Sousa, Subsíd. Estudo Fl. Niassa Port.:**   **(1935).**

Gomes e Sousa A.F. Plantas Menyharthianas.
[Reprinted with new pagination from Bol. Soc. Estudos Colon. Moçamb.
No. 29–32] (1936).   **[F10.71]**

cite as:- **Gomes e Sousa, Pl. Menyharth.:**   **(1936).**

Gomes e Sousa A.F. Essências Florestais de Inhambane.
[Reprinted with new pagination from Moçambique Documentario Trimestral,
22–29] (1943).   **[F10.71]**

cite as:- **Gomes e Sousa, Essenc. Florest. Inhambane:**   **(1943).**

* **Gomes e Sousa A.F.** Dendrologia de Moçambique,
vol. 1 Algumas Madeiras Comerciais, 1950 [with English translation as
Dendrology of Mozambique, vol. 1 Some Commercial Timbers, 1951]. **[F10.71]**

vol. 2 Essências do extremo sul, 1949.   **[not at Kew]**

vol. 3 [does not exist]
vol. 4 Essências da região do Mutuáli, 1958.
vol. 5 Distrito de Manica e Sofala, 1960.

cite as:- **Gomes e Sousa, Dendrol. Moçamb. [vol.]:**   **[year].**

* **Gomes e Sousa A.F.** Dendrologia de Moçambique, Estudo Geral.
Inst. Invest. Agron. Moçamb. vol. 1 (1966), vol. 2 (1967).   **[F10.71]**

cite as:- **Gomes e Sousa, Dendrol. Moçamb. Estudo Geral [vol.]:**   **[year].**

Gonçalves A.E. Catálogo das espécies vegetais vasculares assinaladas na
provínca de Tete, Moçambique.
–I. Pteridophyta, Gymnospermae e Angiospermae (Ranunculaceae-Oxalidaceae).
Garcia de Orta, Sér. Bot. 4,1: 13–92 (1978–79).

–II. Angiospermae (Rutaceae-Leguminosae, excl. Papilionoideae).
Garcia de Orta, Sér. Bot. 4,2: 93–170 (1980).
–III. Angiospermae (Leguminosae, Papilionoideae).
Garcia de Orta, Sér. Bot. 5,1: 59–124 (1981).
–IV. Angiospermae (Chrysobalanaceae-Rubiaceae).
Garcia de Orta, Sér. Bot. 5,2: 139–212 (1982).   **[1.59/16]**

   cite as:- **Gonçalves in Garcia de Orta, Sér. Bot. [vol.]:   [year].**

Henriques J.A. (ed.).   Apontamentos sobre a flora da Zambezia. Exploração
do medico M. Rodrigues de Carvalho.
[Ferns by J.G. Baker, grasses by E. Hackel, sedges by H. Ridley].
Bol. Soc. Brot. 6: 133–144 (1888).
[continued under the title "Contribuições para o conhecimento da flora d'Africa"
in Bol. Soc. Brot. 7: 223–240 (1889) – including also collections from other
Portuguese African territories].   **[1.59/2]**

   cite as:- **[Author] in Henriques in Bol. Soc. Brot. [vol.]:   [year].**

Mendonça F.A. & Torre A.R.   Contribuições para o conhecimento da flora
de Moçambique-I.
Junta Invest. Ultram. Estudos, Ensaios e Documentos. Lisbon. (1950).

   cite as:- **Mendonça & Torre, Contrib. Conhec. Fl. Moçamb. 1:   (1950).**

Mendonça F.A. & others.   Contribuições para o conhecimento da flora de
Moçambique- II.
Junta Invest. Ultram. Estudos, Ensaios e Documentos. 12. Lisbon. (1954).
   [F10.71]

   cite as:- **[Author] in Mendonça, Contrib. Conhec. Fl. Moçamb. 2:   (1954).**

Mogg A.O.D.   An annotated checklist of the flowering plants and ferns of
Inhaca Island, Mozambique.
[in Macnae W. & Kalk M. A Natural History of Inhaca Island, Moçambique:
139–156 (1958)].
Witwatersrand Univ. Press. Johannesburg. 1958.   **[F10.71]**

   cite as:- **Mogg in Macnae & Kalk, Nat. Hist. Inhaca Isl., Moçamb.:   (1958).**

Munday J. & Forbes P.L.   A preliminary check list of the flora of Inhaca
Island, Moçambique: based on the collection of A.O.D. Mogg.
Journ. S. Afr. Bot. 45,1: 1–10 (1979).   **[12/18]**

   cite as:- **Munday & Forbes in Journ. S. Afr. Bot. 45:   (1979).**

Pedro J.G.   Contribuições para o inventário florístico de Moçambique,
1. Pteridofitas-Monocotiledóneas.
Bol. Soc. Estudo Moçamb. 87: 1–53 (1954);
ibid. 2. Dicotiledóneas (Casuarinaceae-Connaraceae). op. cit. 91: 5–30 (1955);
ibid. 3. Leguminosae. op. cit. 92: 5–35 (1955).   **[10.71/3]**

   cite as:- **Pedro in Bol. Soc. Estudo Moçamb. [vol.]:   [year].**

* **Peters W.C.H.**   Reise nach Mossambique, II Abtheilung, Botanik.
Berlin, pages 1–304 (1861), pages 305–end (1864).   **[q F10.71]**

   cite as:- **[Author] in Peters, Reise Mossamb., Bot.:   [year].**

Schinz H.   Plantae Menyharthianae.
[reprinted with new pagination from Denkschr. Math.-Naturwiss. Klasse
Kaiserl. Akad. Wiss. 78: (1905).]   **[q F10.71]**

   cite as:- **Schinz, Pl. Menyharth.:**   **(1905).**

Schinz H. & Junod H.   Zur Kenntnis der Pflanzenwelt der Delagoa-Bay.
[list, in 2 parts and a supplement, of collections made by various collectors,
including Forbes, Junod, Kuntze, Monteiro, Schlechter].

Part 1., Bull. Herb. Boiss. 7: 869–892 (1899).
Part 2., Mem. Herb. Boiss. 10: 25–75 (1900).
Suppl., Bull. Herb. Boiss. Sér. II, vol. 3: 654–662 (1903).   **[1.42/12]**

   cite as:- **Schinz & Junod in Bull. Herb. Boiss. 7:**   **(1899).**
         Schinz & Junod in Mem. Herb. Boiss. 10:   **(1900).**
         Schinz & Junod in Bull. Herb. Boiss. Sér. II, 3:   **(1903).**

*   **Sim T.R.**   Forest Flora & Forest Resources of Portuguese East Africa.
Taylor & Henderson, The Adelphi Press. Aberdeen. (1909).   **[F10.71]**

   cite as:- **Sim, For. Fl. Port. E. Afr.:**   **(1909).**

Sobrinho L.G.   Contribuição para conhecimento da flora de Moçambique.
Bol. Soc. Port. Ci. Nat. 21: 202–214 (1956).
[enumeration of collections made by Cardoso & Serpa Pinto, Barahona e Costa,
Silveira e Sousa].   **[1.59/4]**

   cite as:- **Sobrinho in Bol. Soc. Port. Ci. Nat. 21:**   **(1956).**

ZAMBIA

Astle W.L. The vegetation and soils of Chishinga Ranch, Luapula Province, Zambia.
Kirkia 7: 73–102 (1968). **[10.74/1]**

cite as:- **Astle in Kirkia 7:** **(1968).**

Carr N. Some Common Trees and Shrubs of the Luangwa Valley.
Wildlife Conservation Society of Zambia. 1977.

cite as:- **Carr, Some Common Trees Shrubs Luangwa Valley:** **(1977).**

Fanshawe D.B. Fifty Common Trees of Northern Rhodesia.
Nat. Res. Board, Lusaka. 1962. **[F10.73]**

cite as:- **Fanshawe, Fifty Common Trees N. Rhod.:** **(1962).**

Fanshawe D.B. Check List of Vernacular Names of the Woody Plants of Zambia.
Forest Research Bull., No. 3., Min. Land & Nat. Res., Lusaka. 1965. **[580.4]**

cite as:- **Fanshawe, Check List Vern. Names Woody Pl. Zambia:** **(1965).**

Fanshawe D.B. Fifty Common Trees of Zambia.
Forest Research Department Bull., No. 5., Min. Nat. Res. & Tourism, Lusaka
1968. [different from Fanshawe 1962]. **[F10.73]**

cite as:- **Fanshawe, Fifty Common Trees Zambia:** **(1968).**

* **Fanshawe D.B.** Check List of the Woody Plants of Zambia Showing their Distribution.
Forest Research Bull., No. 22., Min. Land Nat. Res., Lusaka. 1973. **[F10.73]**

cite as:- **Fanshawe, Check List Woody Pl. Zambia Showing Distrib.:** **(1973).**

Fries R.E. Wissenschaftliche Ergebnisse der Schwedischen Rhodesia-Kongo-Expedition 1911–1912, I. Botanische Untersuchungen (1914–1916).
Stockholm, part 1 – Pteridophyta & Choripetalae: 1–184 (1914).
  part 2 – Monocotyledons & Sympetalae: 185–354 (1916).
  Erganzungsheft: 1–135 (1921). **[F10.6]**

cite as:- **Fries, Wiss. Ergebn. Schwed. Rhod.-Kongo-Exped.:** **[year]**

Hutchinson J. A Botanist in Southern Africa.
Gawthorn Ltd., London, 1946.
[index published separately as Lasca Miscellanea, vol. 1, by G.H. Spalding, 1953].
  [F12]

cite as:- **Hutch., Botanist S. Afr.:** **(1946).**

Mitchell B.L. A first list of plants collected in the Kafue National Park.
Puku, vol. 1, 1963. **[F10.73]**

cite as:- **Mitchell in Puku 1:** **(1963).**

Pole Evans I.B. Roadside observations on the vegetation of East and Central Africa.
Bot. Surv. S. Afr. Mem. 22, (1948). map. **[12/20]**

cite as:- **Pole Evans in Bot. Surv. S. Afr. Mem. 22:** **(1948).**

Richards M.A.E. & Morony W.V.   Check List of the Flora of Mbala (Abercorn) & District.   **[F10.73]**

   cite as:- **Richards & Morony, Check List Fl. Mbala & Distr.:**   **(1969).**

Storrs A.E.G.   Know Your Trees.
Forest. Dept., Ndola, 1980.   **[F10.73]**

   cite as:- **Storrs, Know Your Trees:**   **(1980).**

Storrs A. & J.   A Children's Tree Book.
Forest. Dept., Ndola, 1981.   **[F10.73]**

   cite as:- **A. & J. Storrs, Children's Tree Book:**   **(1981).**

Storrs A.E.G.   More About Trees.
Forest. Dept., Ndola. 1982.   **[F10.73]**

   cite as:- **Storrs, More About Trees:**   **(1982).**

Verboom W.C., Fanshawe D.B., Wild H. & Neale K.   Common Weeds of Arable Lands in Zambia.
Min. of Rural Dev. Land Use Services Div. 1973.   **[632.58 (10.73)]**

   cite as:- **Verboom et al., Common Weeds Arable Lands Zambia:**   **(1973).**

Vernon R.   Field Guide to Important Arable Weeds of Zambia.
Dept. Agric., Chilanga. (1983).   **[632.58 (10.73)]**

   cite as:- **Vernon, Field Guide Arable Weeds Zambia:**   **(1983).**

*   **White F.**   Forest Flora of Northern Rhodesia.
Oxford Univ. Press. 1962.   **[F10.73]**

   cite as:- **White, F.F.N.R.:**   **(1962).**

Baker E.G., Exell A.W., & Moore S.   Notes on Dr. R.F. Rand's Rhodesian plants.
Journ. Bot. 64: 301–307 (1926).   **[1.2/86]**

cite as:- **Baker et al. in Journ. Bot. 64:**   **(1926).**

Biegel H.M.   Check List of Ornamental Plants Used in Rhodesian Parks and Gardens.
Rhod. Agric. J. Res., Rep. 3., Min. Agric. [Salisbury] Harare. 1977.

cite as:- **Biegel, Check List Ornam. Pl. Rhod. Parks & Gard.:**   **(1977).**

Biegel H.M. & Mavi S.   A Rhodesian Dictionary of African and English Plant Names. [ed. 2 of H. Wild 1953].
Dept. Res. & Spec. Serv., Min. of Agric. [Salisbury] Harare. (1972).
                                                                [580.4 (10.74)]

cite as:- **Biegel & Mavi, Rhod. Bot. Dict. Pl. Names:**   **(1972).**

Boughey A.S.   A check list of the trees of Southern Rhodesia.
Journ. S. Afr. Bot. 30: 151–176 (1964).   **[12/18]**

cite as:- **Boughey in Journ. S. Afr. Bot. 30:**   **(1964).**

Coates Palgrave O.H.   Trees of Central Africa.
[text by K. Coates Palgrave, not restricted to Zimbabwe, also takes in Malawi & Zambia].
Nat. Publ. Trust, Rhod. & Nyasaland, [Salisbury] Harare.   **[F10.72]**

cite as:- **O. Coates Palgrave, Trees Cent. Afr.:**   **(1956).**

Coates Palgrave K.   Trees of Southern Africa.
[not restricted to Zimbabwe, also takes in Botswana & Mozambique south of the Zambezi R.]
C. Struik Publishers. Cape Town. 1977.   **[F12]**

cite as:- **K. Coates Palgrave, Trees Southern Afr.:**   **(1977).**

Drummond R.B.   The Bundu Book of Trees, Flowers and Grasses.
[ed. 2 of The Bundu Book, 1956].
Longman Rhodesia. 1972.   **[F10.74]**

cite as:- **Drumm., Bundu Book:**   **(1972).**

Drummond R.B.   A List of Trees, Shrubs and Woody Climbers Indigenous or Naturalised in Rhodesia.
Kirkia 10: 229–285 (1975).   **[10.74/1]**

cite as:- **Drumm. in Kirkia 10:**   **(1975).**

Drummond R.B.   Common Trees of the Central Watershed Woodlands of Zimbabwe.
Department of Natural Resources, Harare. 1981.

cite as:- **Drumm., Common Trees C. Watershed Woodl. Zimbabwe:**   **(1981).**

Drummond R.B.   Arable Weeds of Zimbabwe.
Agricultural Research Trust of Zimbabwe, Harare. 1984.

  cite as:- **Drumm., Arable Weeds Zimbabwe:**   **(1984).**

Drummond R.B. & Coates Palgrave K.   Common Trees of the Highveld.
Longman Rhodesia [Salisbury] Harare. 1973.   **[F10.74]**

  cite as:- **Drumm. & K. Coates Palgrave, Common Trees Highveld:**   **(1973).**

Eyles F.   A record of plants collected in Southern Rhodesia.
Trans. Roy. Soc. S. Afr. 5: 273–564 (1916).   **[F10.74]**

  cite as:- **Eyles in Trans. Roy. Soc. S.Afr. 5:**   **(1916).**

Gibbs Russell E.   Keys to vascular aquatic plants in Rhodesia.
Kirkia 10: 411–502 (1977).   **[10.74/1]**

  cite as:- **Gibbs Russell in Kirkia 10:**   **(1977).**

Gibbs L.S.   A contribution to the botany of Southern Rhodesia.
Journ. Linn. Soc., Bot. 37: 425–494 (1906).   **[1.2/97]**

  cite as:- **Gibbs in Journ. Linn. Soc., Bot. 37:**   **(1906).**

Goodier R. & Phipps J.B.   A revised check list of the vascular plants of the
Chimanimani Mountains.
Kirkia 1: 44–66 (1961).   **[10.74/1]**

  cite as:- **Goodier & Phipps in Kirkia 1:**   **(1961).**

Hopkins J.C.F., Bacon A.L., & Gyde L.M.   Common Veld Flowers.
Rhod. Sci. Assn. [Salisbury] Harare. 1940.   **[F10.74]**

  cite as:- **Hopkins et al., Common Veld Fl.:**   **(1940).**

Hoyle J.   Some Flowers of the Bush.
[not restricted to Zimbabwe, also takes in Botswana, Zambia, Malawi &
Mozambique].
Longmans, Green & Co. London. 1953.

  cite as:- **J. Hoyle, Some Fl. Bush:**   **(1953).**

Jacobsen W.B.G.   Check list and discussion of the flora of a portion of the
Lomagundi District, Rhodesia.
Kirkia 9: 139–207 (1973).   **[10.74/1]**

  cite as:- **Jacobsen in Kirkia 9:**   **(1973).**

Linley K. & Baker B.   Flowers of the Veld.
Longman, Rhodesia. 1972.   **[F10.74]**

  cite as:- **Linley & Baker, Fl. Veld:**   **(1972).**

Martineau R.A.S.   Rhodesian Wild Flowers.
Longman Green, London. 1954.   **[F10.74]**

  cite as:- **Martineau, Rhod. Wild Fl.:**   **(1954).**

Monro C.F.H.   Some indigenous trees of Southern Rhodesia.
Proc. Rhod. Sci. Ass. 8,2: 5–123 (1908).

  cite as:- **Monro in Proc. Rhod. Sci. Ass. 8,2:**   **(1908).**

Norlindh T. & Weimarck H. Beiträge zur Kenntnis der Flora von Sud-Rhodesia.
Bot. Notis. 1932–1968. **[F10.74]**

cite as:- **Norlindh & Weimarck in Bot. Notis. [vol.]: [year].**

Pardy A.A. Notes on indigenous trees and shrubs of S. Rhodesia.
[Bound reprints from Rhod. Agric. Journ. (1951–56)]. **[F10.74]**

cite as:- **Pardy in Rhod. Agric. Journ. [vol.]: [year].**

Plowes D.C.H. & Drummond R.B. Wild Flowers of Rhodesia.
Longman Rhodesia. [Salisbury] Harare. 1976. **[F10.72]**

cite as:- **Plowes & Drumm., Wild Fl. Rhod.: (1976).**

Rendle A.B., Baker E.G., Moore S. & Gepp A. A contribution to our knowledge of the flora of Gazaland: being an account of the collections made by C.F.M. Swynnerton.
Journ. Linn. Soc., Bot. 40: 1–245 (1911). **[1.2/97]**

cite as:- **[Author] in Journ. Linn. Soc., Bot. 40: (1911).**

Steedman E.C. A description of some trees, shrubs and lianes of Southern Rhodesia.
[expanded with new pagination from a series of papers in Proc. & Trans. Rhod. Sci. Ass. 1925]. **[F10.74]**

cite as:- **Steedman, Trees, Shrubs & Lianes S. Rhod.: (1933).**

Suessenguth K. & Merxmüller H. A Contribution to the Flora of the Marandellas District, Southern Rhodesia.
[reprinted with new pagination from Proc. & Trans. Rhod. Sci. Ass. 43 (1951)].
(for pagination in original Journal add 74). **[F10.74]**

cite as:- **Suesseng. & Merxm., Contrib. Fl. Marandellas Distr.: (1951).**

Tredgold M.H. & Biegel H.M. Rhodesian Wild Flowers.
Nat. Mus. & Monuments Rhod. [Salisbury] Harare. 1979. **[F10.74]**

cite as:- **Tredgold & Biegel, Rhod. Wild Fl.: (1979).**

Wild H. A Southern Rhodesian Botanical Dictionary of Native and English Plant Names.
Branch of Botany. Harare. 1953. **[580.4 (10.74)]**

cite as:- **Wild, S. Rhod. Bot. Dict. Pl. Names: (1953).**

Wild H. Common Rhodesian Weeds.
Branch of Botany. Harare. 1955. **[F10.74]**

cite as:- **Wild, Common Rhod. Weeds.: (1955).**

Wild H. The endemic species of the Chimanimani Mountains and their significance.
Kirkia 4: 125–157 (1964). **[10.74/1]**

cite as:- **Wild in Kirkia 4: (1964).**

Wild H.   The flora of the Great Dyke of Southern Rhodesia with special
reference to the serpentine soils.
Kirkia 5: 49–86 (1965).   **[10.74/1]**

cite as:- **Wild in Kirkia 5:**   **(1965).**

Wild H.   Geobotanical anomalies in Rhodesia, 1 – The vegetation of copper
bearing soils.
Kirkia 7: 1–71 (1968).   **[10.74/1]**

cite as:- **Wild in Kirkia 7:**   **(1968).**

Wild H.   Geobotanical anomalies in Rhodesia, 3 – The vegetation of nickel
bearing soils.
Kirkia 7, suppl.: 1–62 (1970).   **[F01.73]**

cite as:- **Wild in Kirkia 7, suppl.:**   **(1970).**

Wild. H.   A Guide to the Flora of the Victoria Falls.
In Clark, J. D. (ed.), Victoria Falls Handbook: 121–160 (1952).
Commission for Preservation of Natural and Hist. Monuments and Relics:
Northern Rhodesia. Lusaka.   **[pq F10.73]**

cite as:- **Wild in Clark, Victoria Falls Handbook:**   **1952)**

PART 2

Other African floristic works

* **Brenan J.P.M.**   Check-lists of the Forest Trees and Shrubs of the British Empire No. 5, Tanganyika Territory, Part II (1949).
Imperial Forestry Institute, Oxford. 1949.   **[F10.63]**

   cite as:- **Brenan, Check-list For. Trees Shrubs Tang. Terr.:**   **(1949).**

* **Burtt Davy J.**   A Manual of the Flowering Plants and Ferns of the Transvaal with Swaziland, South Africa.
[incomplete, only parts 1 & 2 published, (1926), (1932).]
Longmans, Green. London. 1926.   **[F12.2]**

   cite as:- **Burtt Davy, Fl. Pl. Ferns Transv. [part]:**   **[year].**

* **Conspectus Florae Angolensis.**
Ministério do Ultramar. Lisboa. 1937 –   **[F10.76]**

   cite as:- **[Author] in C.F.A. [vol.]:**   **[year].**

* **Engler H.G.A.**   Die Pflanzenwelt Ost-Africas, C.
Dietrich Reimer. Berlin. 1895.   **[F10.63]**

   cite as:- **Engl., Pflanzenw. Ost-Afr. C:**   **(1895).**

* **Flora Capensis.**
vols. 1–7 (1860–1925, with supplement to vol. 5, sect. 2 (1933))
      publishers of vols. 1–3 – Hodges Smith & Co. Dublin.
      vols. 4–7 – Lovell Reeve & Co. London.   **[F12]**

   cite as:- **[Author] in F.C. [vol.], [part]:**   **[year].**

* **Flore du Congo Belge et du Ruanda-Urundi.**
Institute National pour L'Etude Agronomique du Congo Belge.
Bruxelles. 1948 –
vols. 1–9 (1948–1960).   **[F10.49]**

   cite as:- **[Author] in F.C.B. [vol.]:**   **[year].**

continued as **Flore du Congo, du Rwanda et du Burundi.**
vol. 10 (1963) and then as family parts to 1971.   **[F10.49]**

   cite as:- **[Author] in F.C.R.B. [10 or family]:**   **[year].**

continued again as **Flore d'Afrique Centrale.**
from 1972 in family parts.   **[F10.49]**

   cite as:- **[Author] in F.A.C. [family]:**   **[year].**

* **Flora of Tropical East Africa.**
[in family parts, (1952 -)]
Crown Agents. London.   **[F10.6]**

   cite as:- **[Author] in F.T.E.A., [family]:**   **[year].**

* **Flora of Southern Africa.**
Botanical Research Institute, Dept. Agric. Pretoria, 1966 –   **[F12]**

   cite as:- **[Author] in F.S.A. [vol.]:**   **[year].**

* **Flora of Tropical Africa.**
  vols. 1–10 (1868–1934).
  L. Reeve & Co. London. [F10]

     cite as:- **[Author] in F.T.A. [vol.], [part]:**   **[year].**

* **Flora of West Tropical Africa.**
  ed. 2, vols. 1–3 (1954–72).
  Crown Agents. London. [F10.2]

     cite as:- **[Author] in F.W.T.A., ed. 2,[vol.]:**   **[year].**

* **Gibbs Russell G.E. et al.**   List of Species of Southern African Plants.
  ed. 2, part 1 in Mem. Bot. Surv. S. Afr., 51. (1985).
       part 2 ibid., 56. (1987). [12/20]

     cite as:- **[Author] in Gibbs Russell et al. in Mem. Bot. Surv. S. Afr., ed. 2,
  [part]:**   **[year].**

* **Hedberg O.**   Afroalpine Vascular Plants: A taxonomic revision.
  Symb. Bot. Upsal. 15: 1–411 (1957). [F10.6]

     cite as:- **Hedberg, Afroalp. Vasc. Pl.:**   **(1957).**

* **Merxmüller H. (ed.)**   Prodromus einer Flora von Südwestafrica.
  [in 166 family parts, (1966–70).]
  J. Cramer. Lehre. [F12.1]

     cite as:- **[Author] in Merxm. Prodr. Fl. SW. Afr. [family part]:**   **[year].**

* **Ross J.**   Flora of Natal.
  Dept. Agric. Techn. Serv., Bot. Surv. Mem., 39 (1973). [F12.5]

     cite as:- **Ross, Fl. Natal.:**   **(1973).**

PART 3

Floristic works chronologically arranged

1850–55	Bertoloni, Ill. Piante Mozamb. [Diss.]: [year]	MOZ
1860–1933	[Author] in F.C. [**vol.**]: [year].	EXT
1861–64	[Author] in Peters, Reise Mossamb., Bot.: [year].	MOZ
1868–1934	[Author] in F.T.A. [**vol.**]: [year].	EXT
1888–89	[Author] in Henriques in Bol. Soc. Brot. [**vol.**]: [year]	MOZ
1894	[Author] in Trans. Linn. Soc. ser. 2, Bot. **4**: (1894).	MAL
1895	Engl., Pflanzenw. Ost-Afr. **C**: (1895).	EXT
1897	Burkill in Johnston, Brit. Cent. Afr.: (1897).	MAL
1899–1903	Schinz & Junod in Bull. Herb. Boiss. [**vol.**]: [year]	MOZ
1904	Passarge, Die Kalahari: (1904).	BOT
1905	Schinz, Pl. Menyharth.: (1905).	MOZ
1906	Gibbs in Journ. Linn. Soc., Bot. **37**: (1906).	ZIM
1908	Monro in Proc. Rhod. Sci. Ass. **8**,2: (1908).	ZIM
1909	Sim, For. Fl. Port. E. Afr.: (1909).	MOZ
1909	N.E. Br. in Bull. Misc. Inf., Kew **1909**: (1909).	BOT
1911	Rendle et al. in Journ. Linn. Soc., Bot. **40**: (1911).	ZIM
1914–21	Fries, Wiss. Ergebn. Schwed. Rhod.-Kongo-Exped.: [year].	ZAM
1916	Eyles in Trans. Roy. Soc. S.Afr. **5**: (1916).	ZIM
1926	Baker et al. in Journ. Bot. **64**: (1926).	ZIM
1926–32	Burtt Davy, Fl. Pl. Ferns Transv. [part]: [year].	EXT
1927–36	[Author] in F.W.T.A. ed. 1, [**vol.**]: [year].	EXT
1932–68	Norlindh & Weimarck in Bot. Notis. [**vol.**]: [year].	ZIM
1933	Steedman, Trees, Shrubs & Lianes S. Rhod.: (1933).	ZIM

1935	Gomes e Sousa, Subsid. Estudo Fl. Niassa Port.: (1935).	MOZ
1935	Bremek. & Oberm. in Ann. Transv. Mus. **16**: (1935).	BOT
1936	Gomes e Sousa, Pl. Menyharth.: (1936).	MOZ
1936	Burtt Davy & Hoyle, Check-list For. Trees Shrubs Brit. Emp. 2, Nyasaland: (1936).	MAL
1937–	[Author] in C.F.A. [**vol.**]: [year].	EXT
1940	Hopkins et al., Common Veld Fl.: (1940).	ZIM
1943	Gomes e Sousa, Essenc. Florest. Inhambane: (1943).	MOZ
1946	Hutch., Botanist S. Afr.: (1946).	ZAM
1949–60	Gomes e Sousa, Dendrol. Moçamb. [**vol.**]: [year]	MOZ
1948–60	[Author] in F.C.B.: [year].	EXT
1948	O.B. Mill., Check-list For. Trees Shrubs Bech. Prot.: (1948).	BOT
1948	Pole Evans in Bot. Surv. S. Afr. Mem. 21: (1948).	BOT
1948	Pole Evans, Roadside Obs. Veg. E. & C. Afr.: (1948).	ZAM
1949	Brenan, Check-list For. Trees Shrubs Tang. Terr.: (1949).	EXT
1949–51	Gomes e Sousa, Dendrol. Moçamb. [**vol.**]: [year].	MOZ
1950	Mendonça & Torre, Contrib. Conhec. Fl. Moçamb. **1**: (1950).	MOZ
1951–56	Pardy in Rhod. Agric. Journ. [**vol.**]: [year].	ZIM
1951	Suesseng. & Merxm., Contrib. Fl. Marandellas Distr.: (1951).	ZIM
1952–	[Author] in F.T.E.A., [family]: [year].	EXT
1952–53	O.B. Mill. in Journ. S. Afr. Bot. [**vol.**]: [year].	BOT
1952	Wiehe, Cat. Fl. Pl. Herb. Pl. Path. Zomba: (1952).	MAL
1952	Wild in Victoria Falls Handbook: (1952).	ZIM

1953–54	[Author] in Mem. N.Y. Bot. Gard. [**vol.**], [part]: [year].	MAL
1953	Brass in Mem. N.Y. Bot. Gard. **8**,3: (1953).	MAL
1953	J.Hoyle, Some Fl. Bush: (1953).	MAL
1953	Wild, S. Rhod. Bot. Dict. Pl. Names: (1953).	ZIM
1954–72	[Author] in F.W.T.A. ed. 2, [**vol.**]: year].	EXT
1954	Martineau, Rhod. Wild Fl.: (1954).	ZIM
1954	[Author] in Mendonça, Contrib. Conhec. Fl. Moçamb. **2**: (1954).	MOZ
1954–55	Pedro in Bol. Soc. Estudo Moçamb. [**vol.**]: [year].	MOZ
1955	Wild, Common Rhod. Weeds: (1955).	ZIM
1956	O. Coates Palgrave, Trees Cent. Afr.: (1956).	ZIM
1956	Sobrinho in Bol. Soc. Port. Ci. Nat. **21**: (1956).	MOZ
1957	Chapman, Indig. Conifers Nyasal.: (1957).	MAL
1957	Hedberg, Afroalp. Vasc. Pl.: (1957).	EXT
1957	Willan & McQueen, Exotic Trees Nyasaland: (1957).	MAL
1958	Mogg in Macnae & Kalk, Nat. Hist. Inhaca Isl., Moçamb.: (1958).	MOZ
1958	Topham, Check List For. Trees Shrubs Nyasaland Prot.: (1958).	MAL
1959	Leistner in Koedoe **2**: (1959).	BOT
1961	Goodier & Phipps in Kirkia **1**: (1961).	ZIM
1962	Chapman, Veg. Mlanje Mt.: (1962).	MAL
1962	Fanshawe, Fifty Common Trees Northern Rhodesia: (1962).	ZAM
1962	White, F.F.N.R.: (1962).	ZAM
1963–71	[Author] in F.C.R.B. [**10** or family]: [year].	EXT
1963	Mitchell in Puku **1**: (1963).	ZAM

1964	Boughey in Journ. S. Afr. Bot. **30**: (1964).	ZIM
1964	Wild in Kirkia **4**: (1964).	ZIM
1965	Fanshawe, Check List Vern. Names Woody Pl. Zambia: (1965).	ZAM
1965	Wild in Kirkia **5**: (1965).	ZIM
1966–	[Author] in F.S.A. [**vol.**]: [year].	EXT
1966–	[Author] in Merxm. Prodr. Fl. SW. Afr. [fam. part]: [year].	EXT
1966–67	Gomes e Sousa, Dendrol. Moçamb. Estudo Geral [**vol.**]: [year].	MOZ
1968	Amico & Bavazzano in Webbia **23**: (1968).	MOZ
1968	Astle in Kirkia **7**: (1968).	ZIM
1968	Binns, First Check List Herb. Fl. Malawi: (1968).	MAL
1968	Fanshawe, Fifty Common Trees Zambia: (1968).	ZAM
1968	Wild in Kirkia **7**: (1968).	ZIM
1969	Richards & Morony, Check List Fl. Mbala & Distr.: (1969).	ZAM
1970	Chapman & White, Evergreen For. Malawi: (1970).	MAL
1970	Wild in Kirkia **7**, suppl.: (1970).	ZIM
1972	Biegel & Mavi, Rhod. Bot. Dict. Pl. Names: (1972).	ZIM
1972	Binns, Dict. Pl. Names Malawi: (1972).	MAL
1972	Drumm., Bundu Book: (1972).	ZIM
1972–	[Author] in F.A.C. [family]: [year].	EXT
1972	Linley & Baker, Fl. Veld: (1972).	ZIM
1972	Ross, Fl. Natal.: (1972).	EXT
1973	Brummitt in Wye Coll. Malawi Proj. Rep.: (1973).	MAL
1973	Drumm. & K. Coates Palgrave, Common Trees Highveld: (1973).	ZIM

1973	Fanshawe, Check List Woody Pl. Zambia Showing Distrib.: (1973).	ZAM
1973	Jacobsen in Kirkia **9**: (1973).	ZIM
1973	Ross, Fl. Natal.: (1973).	EXT
1973	Verboom et al., Common Weeds Arable Lands Zambia: (1973).	ZAM
1974–75	Banda & Seyani, Veg. Surv. Sanjika Hill [**vol.**]: [year].	MAL
1975	Banda & Patel, Veg. Surv. Chilunga Estate, Chancellor College, Zomba: (1975).	MAL
1975	Drumm. in Kirkia **10**: (1975).	ZIM
1975	Moriarty, Wild Fl. Malawi: (1975).	MAL
1975	Williamson, Useful Pl. Malawi: (1975).	MAL
1976	Plowes & Drumm., Wild Fl. Rhod.: (1976).	ZIM
1977	Biegel, Check List Ornam. Pl. Rhod. Parks & Gard.: (1977).	ZIM
1977	Carr, Some Common Trees and Shrubs Luangwa Valley: (1977).	ZAM
1977	K. Coates Palgrave, Trees Southern Afr.: (1977).	ZIM
1977	Gibbs Russell in Kirkia **10**: (1977).	ZIM
1977	Howard-Williams in Kirkia **10**: (1977).	MAL
1978–82	Gonçalves in Garcia de Orta, Sér. Bot. [**vol.**]: [year].	MOZ
1979	Munday & Forbes in Journ. S. Afr. Bot. **45**: (1979).	MOZ
1979	Tredgold & Biegel, Rhod. Wild Fl.: (1979).	ZIM
1980	Hall-Martin & Drumm. in Kirkia **12**: (1980).	MAL
1980	Storrs, Know Your Trees: (1980).	ZAM
1981	Berrie in Luso: (1981).	MAL
1981	Drumm., Common Trees C. Watershed Woodl. Zimbabwe: (1981).	ZIM

| 1981 | A. & J. Storrs, Children's Tree Book: (1981). | ZAM |

1981   Woollard, Vegetative Key Woody Pl. SE. Botswana: BOT
          (1981).

1982   Banda & Salubeni, Veg. Surv. Kamuzu Academy   MAL
          Mtunthama, Kasungu: (1982).

1982   Pullinger & Kitchin, Trees Malawi: (1982).   MAL

1982   Storrs, More About Trees: (1982).   ZAM

1983   Balsinhas in Bothalia **14**: (1983).   MOZ

1983   Vernon, Field Guide Imp. Arable Weeds Zambia: ZAM
          (1983).

1984   Sherry & Ridgeway, Field Guide Lengwe Nat. Park: MAL
          (1984).

1984   Drumm., Arable Weeds Zimbabwe: (1984).   ZIM

1985–   [Author] in Gibbs Russell et al. in Mem. Bot. Surv.   EXT
          S. Afr., ed. 2, [part]: [year].

1986   Banda & Morris, Common Weeds Malawi: (1986).   MAL

1986   Barnes & Turton, List Fl. Pl. Botswana at Nat.   BOT
          Mus., Sebele & Univ. Bot.: (1986).

1986   Grignon & Johnsen, Towards Check List Vasc.   BOT
          Pl. Botswana: (1986).

1987   Seyani et al., Plant Sp. Lilongwe Nat. Sanctuary: MAL
          (1987).

1988   Blomb.-Ermat. & Turton, Some Fl. Pl. SE. Botswana: BOT
          (1988).

1988   Chapman in Nyala **12**: (1988).   MAL

1988   Dowsett-Lemaire in Bull. Jard. Bot. Nat. Belg. **58**:   MAL
          (1988).

1989   Blackmore et al. in Kirkia **13**: (1989).   MAL

1989   Shorter, Introd. Common Trees Malawi: (1989).   MAL